Investigating Science with Nails

To my father,
who has made nails a part of his life
. . . and to my mother
who encourages him to use them.

Investigating Science
with Nails

by Laurence B. White, Jr.

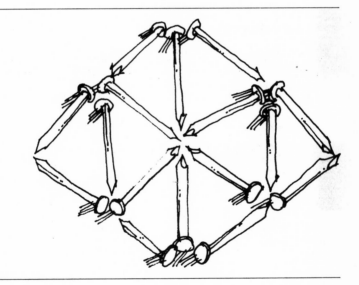

 Addison-Wesley Publishing Company

Reading, Massachusetts

Books by Laurence B. White, Jr.
Investigating Science with Coins
Investigating Science with Rubber Bands
Investigating Science with Paper
Investigating Science with Nails
So You Want to be a Magician?

 **An Addisonian Press Book**

Text copyright © 1970 by Laurence B. White, Jr.
Illustrations copyright © 1970 by Addison-Wesley Publishing Company, Inc.

Addison-Wesley Publishing Company, Inc.
Reading, Massachusetts 01867

Library of Congress Catalog Card Number 74-124880

Printed in the United States of America
Second Printing
ISBN: 0-201-08651-4

Table of Contents

1

Another Way of Looking

Everyone lives in a different world. This is what makes life so interesting. It is why some people become butchers, others become barbers, or teachers, or scientists. We all learn about the things that interest us most and we build our own personal world around them.

Of course, everyone knows a little about science, just as a scientist wants to know a little about what a barber or butcher does. Everyone probably knows just enough about the jobs of others to satisfy his curiosity. You are free to build your own world, just the way you want it.

Not everyone wants to become a scientist, but it is well worthwhile to take a look at the world in the way a scientist does. We all have our own way of looking, and you may find the scientist's way is the most unusual one yet.

This is a book about nails, but it is not a carpenter's book that shows you how to construct things with them. A scientist would look at nails in a different way than a barber, butcher, or office worker would. A scientist would see the nail as a special weight and size. He would realize that it could be used for certain things better than other objects. He would wonder what else, besides building things, nails could be used for. So let's find out how a scientist would look at nails.

To get you started, here is an investigation that uses a nail, but does not require a hammer, a board, or any carpentry skill at all. It does, however, require you to begin thinking like a scientist.

Silver Nail . . . Black Fingers

Things Needed:
A large nail
A candle
Matches
A glassful of water

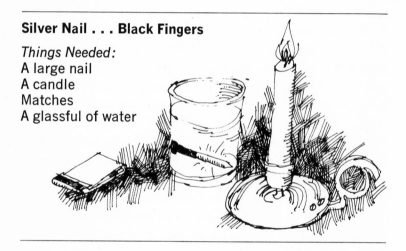

Light the candle. Hold the nail over the flame. The nail should not be in the flame, but in the smoke above it. The soot from the smoke will cover the nail with a thick black layer. Turn the nail making sure that all of it is blackened. You only have to cover half the nail (away from your fingers) with soot.

Drop the nail into the glass of water and observe it through the sides of the glass. You will find that the end not covered with soot appears dull, but the soot-covered end glistens and appears to shine with a silver-colored brightness. The blackened end will appear almost jewel-like under water, but if you reach under water and touch it with your finger, you will find your finger has black soot on it when you take it out of the water.

How can this happen? Why should such a black object appear so shiny and beautiful just because it is under water?

This certainly is "another way of looking" at a nail. A carpenter might have difficulty explaining it, but a scientist could do it easily.

The scientist would explain that certain chemicals have an *affinity* for gases. Affinity simply means an attraction for something else. The chemicals making up the black soot attract air, and cause a layer of air to cling to the nail. The scientist would use the word *adsorption* to describe what was happening. Soot, which is actually the chemical element *carbon*, adsorbs air. It adsorbs a layer of air so completely that as the blackened nail is placed under water, the air remains on its surface and goes underwater with the soot. The light strikes this thin air layer and reflects back to your eye. Materials that reflect light back appear white or silvery. You are not seeing the nail or the soot, you are actually "seeing" the air.

Perhaps this explanation satisfies you. A scientist though, might still have other questions: Why do certain chemicals adsorb gases? How do things reflect light? Why is the reflected light silvery-colored? Why does the soot stick to the nail? How does the candle make soot? There are as many questions yet to be asked as there are pages in this book.

Every one of these explanations and questions gives you "another way of looking" and that, more than anything else, is what you can discover from this book.

2

Get To the Point

How do you like this description of a nail?

"A nail is a crazy piece of metal that is headed in one direction but pointed in the other."

Actually this is not as silly a definition as it first sounds. A nail is made of metal, and it does have a head on one end and a point on the other.

Naturally, the head is flat so you can hit it better with a hammer, and the point is sharp so it goes into the wood easier, but these ends can be used for many other things as well. In this chapter you will be using these heads and points to learn more about nails, yourself, weight, and the science behind nail-pounding.

Let's start with a puzzle that depends on the very fact that nails have two ends with different shapes. Would you believe that these same heads and points can result in an observation test that may completely confuse a friend of yours? Then you can "get to the point" by "turning your heads".

11

HEADS UP AND POINT DOWN

Almost everyone enjoys a puzzle. It is fun to find a solution to something that appears quite impossible. The only thing wrong with most puzzles is that usually the fun is gone once you learn, or are shown, how to solve them. A puzzle is only a puzzle when your mind cannot see the solution.

If you agree, then you might enjoy this double puzzle. First you must learn how to do it, then, with just one little trick, you can puzzle your friends even more. First start by attempting to solve it yourself:

Turning Your Heads

Things Needed:
Three nails with flat heads

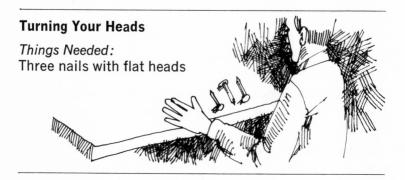

Lay the nails in a row on the table. Adjust them so that the head of the center nail is away from you and the heads of the outside nails are toward you.

The object of the puzzle is to turn the nails so that all three *heads* are facing *away* from you. Before you begin, however, there are two rules you must follow. These rules, you will discover, make the puzzle a bit more difficult than it first sounds. First, you must always turn two nails at the same time. This counts as one turn. Second, you must make exactly three turns, no more, no less. You may turn any two nails you prefer on each turn, but choose carefully because, after three turns you must have all three heads facing away from you.

Now try it, before you read any further!

If you have spent a few minutes with this puzzle you probably discovered that it was interesting, but not terribly difficult. Almost everyone solves the puzzle after a few attempts. (By the way, there are three ways to solve it.)

If, by chance, you haven't succeeded, here is how to do way number one:

Think of the nail to your left as being nail number 1. Nail number 2 is in the middle, and the nail to your right is number 3. On your first turn, turn nails number 1 and 2. Next, (second turn) turn 1 and 3. For your last turn, turn nails 1 and 2 once again to bring all the heads away from you.

To do way number two, simply turn nails number 1 and 3 three times. And finally, to do way number three, turn nails 2 and 3 the first time. Then nails 1 and 3 the second time, and nails 2 and 3 for the final turn.

You might wonder what this puzzle has to do with science. Once you understand how to solve the problem the fun is not gone. You can now use the puzzle to conduct a simple observation test on a friend.

Much of what your mind knows has entered through your eyes. Some people call eyes the "windows of the mind". Sight is a sense, and this sense together with the senses of smell, hearing, taste, and touch supply our minds with everything we know about the world around us. It's easy to understand, then, that your mind is only as good as the information supplied by your senses. If your senses make a mistake in telling your brain something, you will make the same mistake when you try to remember it later. Another way your brain might make a mistake is by not watching carefully enough. If your eyes do not give your brain certain information, then your brain certainly can never remember it. Supplying your brain with facts, that you see from your eyes, is called observing. Most people are not careful observers and you can use this nail-turning puzzle to prove it.

Show your friend the puzzle with the three nails. Lay the three nails on the table as you did before and carefully explain the two rules to him (his ears will give this information to his brain). Now, show him how to solve the puzzle, using way number one. Do this before he has any opportunity to try it. Remember, your investigation is not only to see if he can find the solution, but how well he observes the problem. As you solve the puzzle, and do the turns, your friend's eyes are feeding this new information to his brain. He is observing. Now, pick up all the nails in your hand and ask him if he would like to try it.

Does your friend understand the puzzle? You certainly have supplied his brain with all of the information it should require. His ears have heard you explain it, and his eyes have actually seen you do it. He knows that it is not impossible. Your friend will probably be anxious to try it just to prove that he is as clever as you.

Now, and read this carefully, here is the test you will

perform to see if your friend really is a good observer: Lay the three nails on the table in front of him *but*, turn them so that the center nail has its head facing toward him and the two outer nails have their points facing toward him. This is just the opposite arrangement that you used.

If your friend notices this different arrangement he certainly is a careful observer. (If he doesn't notice it he will quickly learn that he is not.) Let him try to complete the puzzle. He will find that it is impossible to make all the heads face away when he starts with this wrong arrangement. He will probably become irritated with himself for being unable to remember what you did, and it will become an unsolvable puzzle for him.

Most science experiments are puzzles when we start. So to solve them properly we must be careful observers.

A TOUCHY POINT

Someone once defined skin as a "bag that holds you together". To a scientist, however, skin is much more important than just a "bag". The skin is an *organ*, and like other organs (such as the heart, kidneys, brain, etc.) skin performs several functions. Probably the most important function is that of perspiring.

Sweat glands, located in the skin, accumulate waste materials (excess water, carbon dioxide and dissolved salts) and eliminate these wastes through the tiny openings called pores. Once outside the body, these wastes evaporate into the air.

But skin does more than that. Imagine a square, three-fourths of an inch on each side, drawn on the back

of your hand. The square
would be this size:

If you had a very powerful microscope, and you
scraped away layer after layer of skin, right down to the
flesh underneath, as you examined it carefully, in that
tiny square you might find. . . .

9 million cells
13 yards of nerves
9 feet of blood vessels
9000 nerve endings
300 sweat glands
4 oil glands
6 cold sensors
36 heat sensors
75 pressure sensors
600 pain sensors
and, 30 hairs

Of course these numbers are not exact. Some parts
of your skin have more or less of some of the items, but
the quantities are reasonable for a three-fourths inch
square of skin on the back of your hand.

You already know your five senses: touch, taste,
sight, smell and hearing. These are called *major* senses.
You also have some *minor* senses. For example, you
know when something feels hot or cold. This can be
called your temperature sense. You can investigate this
sense, and others, with a sharp pointed nail:

Point to Your Senses

Things Needed:
A sharp pointed nail

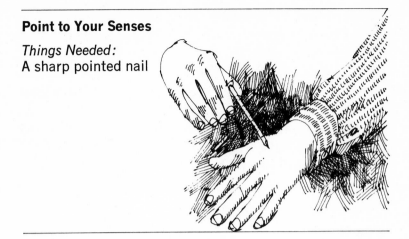

   Hold the head end of the nail in one hand and gently
touch the point of the nail to the skin on the back of
your other hand. If you feel the sharp point of the nail
you have touched one of the pain sensors (the most
common kind of sensors). If the point feels dull, but
you can feel it pressing, then the point is probably
resting on one of the pressure sensors (they are the
second most common kind). Now, move the nail just
a tiny distance and press it down again. You may feel
a different sensation. The most surprising will come
when you touch a cold sensor, and the nail point
suddenly feels quite cool.
   Now find a hair in your skin. Wiggle it back and
forth with the point of the nail. You will discover a
new sensor, which is actually part of your touch sense.
It is called, quite scientifically, a tickling sense!
   When the doctor gives you a "shot", he generally
uses your arm as his target. But suppose he decided to

use your finger instead of your arm? Just thinking about that may make you shiver because you know that your fingers are very sensitive to pain.

Different parts of your skin have different jobs to do. Our fingers do most of our feeling so it is not surprising to learn that they have the greatest number of touch and pain sensors. Other parts of your skin lack them altogether, and in some places, you can't even tell the point of a nail from the head!

Head or Point?

Things Needed:
A small
   flat-headed nail
A friend

This is an investigation you can try on a friend, and have your friend try on you. Start with yourself. Give your friend the nail and close your eyes. With your eyes closed ask your friend to touch various parts of your skin with *either* the head or the point of the nail. You must guess whether it is the head or the point that touches you. Have him start by touching your fingertips. You will probably guess correctly every time. Ask him to try the back of your hand, then your arm, parts of your face, legs, and so on. Be sure that he presses down very lightly each time, so that the nail touches only a few sensors. Don't be surprised if you discover yourself making many errors once he starts touching places other than your hands.

Your friend may laugh when you cannot distinguish between a big flat head and a sharp point, but try the same investigation on him. You may find he cannot tell the difference between the two ends either.

A BED OF NAILS

Probably nobody "gets to the point" of nails more dramatically than the fakir magicians of India. They rest on boards covered with the sharp points of hundreds of nails.

There is no magic at all in this trick. You could even do it if you really wanted to. It would not require much practice, nor would it require a very tough skin, but it certainly might help ease your mind if you knew a bit of science. A scientist would tell you that this ancient trick actually depends on the scientific principle of *distribution of weight*.

Here's how it works: Think of a table with four legs. Each leg supports one-fourth of the total weight (if the weight is equally distributed on each leg). Now think of a camera tripod. It has three legs, therefore each leg would support one-third of the total weight on the tripod. Think of yourself. You have two legs, and each one must hold up one-half the total weight of your body.

Of course, if you stand on one foot all of your weight will be supported by just one leg. This, then, is the idea of weight distribution. Each leg helps support the total weight.

If you are in the mood to do a lot of nail pounding you can investigate this fakir's trick. For safety's sake though . . . please do it the opposite way.

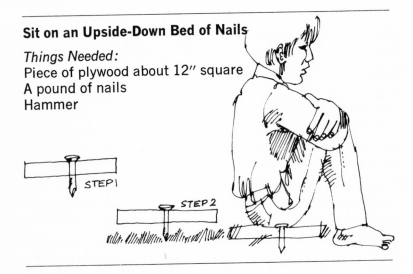

Sit on an Upside-Down Bed of Nails

Things Needed:
Piece of plywood about 12″ square
A pound of nails
Hammer

Do this investigation outdoors. You will require a fairly soft patch of earth or grass to work on.

Step One: Drive one nail through the center of the board. *Step Two:* Place the board and nail on the ground (the pointed end of the nail facing toward the ground). *Step Three:* Sit on the board. Your weight will force the nail into the ground and the board flat on the ground. When you sit on the board it acts like a table, and the single nail acts like one "leg". Since the nail (leg) must support all of your weight, it is driven into the soft ground.

It's easy to see why you should do this investigation upside-down! If you tried to sit directly on the pointed end of a single nail your weight certainly would push it deep into your flesh. For this reason it is a very foolish and dangerous trick to place a tack on a chair seat and wait until some friend sits down on it.

Now drive three more nails into the board at different places. Place the board and nails on the ground (nail points down), and again, sit on the board. The four nails will probably sink into the ground, but you will notice that they seem to hold your weight better. Your total weight is distributed among the four of them.

Now drive a number of nails into the board and try the investigation again. How many nails must you use before the board will support your weight without sinking deeply into the ground?

Suppose you could sit evenly over all the nails at the same time, making sure that each nail supported an equal portion of your weight. If your board had 100 nails, and you weighed 100 pounds, each nail would support one pound. If your board had 1600 nails, each nail would support only one ounce of weight.

With a board containing many hundreds of nails you can see how it would be possible, with the points facing up, for a fakir to lay his entire weight down on his "bed". Each nail supports only a very small part of his weight. No single nail supports enough weight to allow it to penetrate his skin, or hurt him. Naturally, this bed is not very comfortable, and the fakir lies on it for a show, and never for actual sleeping.

Although it would be entirely possible for you to do his trick, should you have a bed containing hundreds of nails, we "science fakirs" should always do it upside-down. That way the only thing that ends up with holes in it is the ground!

WHEN WE WANT WEIGHT

A hammer appears to be a simple tool, and we all know it works, but rarely wonder how.

A nail itself is actually a one-legged table. Any weight you apply to its head must be supported by the body of the nail. Therefore, the force of a hammer striking its head is transmitted to the nail, which is driven into the piece of wood.

How much force must a nail receive? Naturally, the more the better, but just how much force can a hammer apply to a nail? This depends on two things that you can investigate.

How Heavy the Hammer

Things Needed:
Two different hammers
  One small and light
  one heavy and large
A piece of board
Two nails

Actually, this is rather an obvious investigation. Simply try driving a nail with each of the two hammers. You will notice that it is far easier to pound a nail into a board quickly with the heavy hammer. Why? Driving a nail requires you to do some work and use some of your energy. It is harder work to lift and swing the heavier hammer, but in return, the hammer contains more energy which will apply a greater force to the nail.

To use a heavy hammer you must do a lot of work in a very short time. You can get the same force from a lighter hammer by doing the same amount of work, but taking more time to do it. The handle of the hammer helps you do this very thing.

How Long Is The Handle

Things Needed:
Hammer, any size
Nails
Piece of board

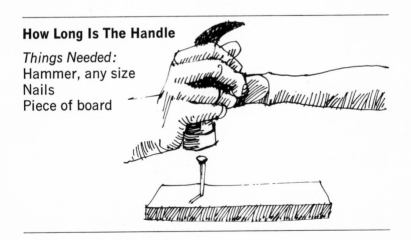

Hold the *head* of the hammer in your hand and try to pound a nail into the board. You will find this a difficult task. To make the task a little easier, hold the hammer handle halfway down and pound another nail. Finally, hold the handle, in the usual manner, toward the end away from the head. This is the easiest way.

It requires more work on your part to make a hammer move more quickly. The faster the hammer head is traveling when it meets the nail, the more force it can apply.

Would you prefer to catch a bowling ball that was tossed to you slowly, or a baseball that was thrown hard? They would both hit you with considerable force wouldn't they? For the same reason a light hammer, moving very quickly, can drive a nail just about as well as a heavy hammer moving more slowly. The hammer handle helps in this investigation, but if you have a very heavy hammer with a long handle, it still might be impossible for you to drive a nail. This investigation will show you why.

Investigate the Starting Distance

Things Needed:
Hammer, any size
Nails
Board

Here's a challenge for you: Try to drive a nail into the board without ever raising the hammer more than one inch over the head of the nail. You may hold the hammer in any way you wish, but never raise it above the nailhead more than an inch. Perhaps you will succeed, but you will probably give it up as impossible.

When the hammer is only a short distance above the nail, you are unable to start it moving quickly before it meets the nail. The hammer does not contain much energy because you do not have time to transfer much of your energy to the hammer. If, however, you start higher you will have more time to give the hammer more energy. This distance is called the *starting distance*. The longer the starting distance the more energy you can put to work for you. Can you see how starting distance would help you hit a baseball, kick a football, or shoot an arrow? Which would you think is most important in nail driving: the weight of the hammer head, the length of the handle, or the starting distance?

Actually, all three of these ideas work together: the weight of the hammer, the length of the handle and the starting distance. Driving a nail requires energy that the hammer gets from you. The weight of the hammer, the handle, and how you use it all help put a great deal of your energy to work in the most efficient way.

If this is true the best hammer would perhaps weigh 100 pounds, but it is easy to understand why they're not made that heavy. It would require a giant carpenter to lift one. Also, the best hammer should be one with the longest handle. You could get an awful lot of force from a hammer with a ten foot handle. But they don't make hammers with handles ten feet long.

A DULL POINT

Why is the point of a nail sharp? This sounds like a foolish question, but be careful before you answer because the question is not as silly as it sounds. Do you know that not all nails have sharp points? If you go to a hardware store and ask for some "cut" nails you will receive a nail which has flat sides, a flat head, and a flat point. What is the advantage to a flat pointed nail? What is the advantage to a nail having a sharp point?

   Most people would guess that a sharp nail would be better than a blunt one simply because it would be easier to drive into the wood. Surprisingly, this is not always true.

Hammer and Nail Surprise

Things Needed:
Two nails the same size
Hammer
Thin board
File
Two books, same size
A worktable

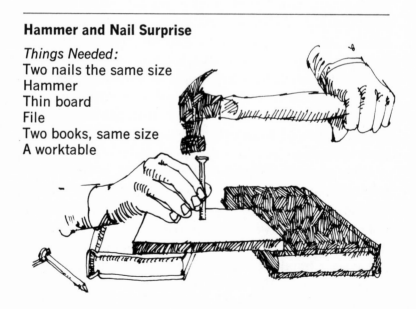

Place the two books on the worktable with the board on top to form a bridge. Flatten the point of one of the nails with the file. Be sure to make it blunt and perfectly flat on the end. Now, drive each of the nails into the piece of wood, all the way through.

You will discover that the blunted nail goes through the board as easily, and perhaps even more easily, than the sharp nail. This will probably surprise you if you've never done it before.

Think, for a moment, about what happened when each of the nails enters the wood. The sharp nail pushes the fibres of the wood aside, like the sharp point of a boat pushes the water aside as it moves forward. Two forces are working against each other as the nail "knifes" its way into the wood; the nail forcing the fibres apart, and the springiness of the fibres pushing back against the nail. Actually the fibres pushing against the nail makes the job of driving the nail harder.

The blunt nail, however, punches a hole through the board. The flat end of the nail pushes against the wood that is directly underneath it. You might even find a little cylinder of wood underneath the board after the nail emerges.

Why might a carpenter choose blunt pointed nails for certain building projects? There are many differences between the two kinds of nails, and perhaps studying the differences a bit more might help you answer that question.

Here's one idea: As the sharp nail knifes into the wood it pushes the fibres aside. If the board were narrow there is a chance that the nail may split the board down the middle. Surprisingly, sharp pointed nails are more apt to split the board than a blunt one.

You might begin to wonder why any nails are sharp pointed. Aren't blunt nails actually the best kind? If they were I suspect there would be more blunt nails than sharp nails for sale in the hardware shop. Perhaps you had better investigate further.

Pulling Your Nails

Things Needed:
Board with nails from
   last investigation
Claw hammer

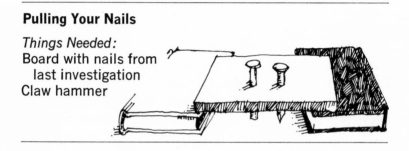

With your fingers, take hold of the head of the blunt pointed nail. You will find that it wiggles easily. The hole the blunt nail has made in passing through the board is made even larger by your wiggling. Now try wiggling the sharp pointed nail. The fibres pressing against this nail will grip it tightly and make wiggling much more difficult.

Which nail is the easier to remove from the board? You can probably, after a little wiggling, pull the blunt nail out with your fingers. You will probably have to use the claw of the hammer to draw out the sharp pointed nail. It is obvious that the wood fibres are very important for holding a nail in place.

Seeing The Difference

Things Needed:
Board from last investigation,
    with nails removed

In your investigations, so far, you have driven two nails, both the same size, into a piece of board and you have removed them. Both holes should be the same size! But they're not! Look carefully at them.

The hole made by the blunt nail will be quite round, but the hole made by the sharp nail will be smaller and flatter. The fibres were able to spring back into the hole left by the sharp nail, but a hole was punched out by the blunt nail.

There is another difference between the two that you probably have already observed. Turn the board over and look at the holes from the underside. The wood around the hole left by the blunted nail is splintered out, but this is not true of the hole left by the sharp nail.

Both the sharp and the blunt nails have worked well for you, but there are many differences between the two. Carpenters have used, and still use, both kinds for different jobs. A carpenter would know where each kind works best, and a scientist could make a good guess from these investigations.

3

Now Meet Heat

You already know what heat is if you have ever warmed your hands by a fire or burned your finger on a hot stove. But if you were asked to give a definition for the word *heat*, what would your definition be?

Before you answer, here are three facts that may confuse you and make you think a bit harder before answering: An ice cube contains heat! A cold object may actually be quite hot! Heat can make you feel warm, or cold!

Do these facts raise another question? *What is cold?* It is practically impossible to talk about heat without mentioning cold and both of these words are equally difficult to define.

These questions are not really meant to confuse you. They simply remind you that such commonly used words as heat, cold and temperature are really worth studying. You'll find them much easier to understand by doing the investigations in this chapter.

If we wire a light bulb to the battery, the electrical energy is changed to two other kinds of energy; light and heat.

If you would like to remember all of the most common forms of energy just remember the word SCREAM. Each letter in the word scream will remind you of the first letter of a form of energy.

Sound
Chemical
Radiant (heat and light)
Electrical
Atomic
Mechanical

Each of them can be changed into another form of energy—very easily. For example, when you scream, the sound energy is changed to heat energy when it strikes the walls of a room. Inside your friend's ears it changes to mechanical energy by causing his eardrums to vibrate.

The tremendous heat energy produced by atomic power can cause water to become steam. This steam can be used to turn giant electrical generators which can provide enough electrical energy to light all the buildings in a city.

The examples of changing one form of energy to another are so common you could easily fill a book much larger than this one with just such examples. But instead of just listing more examples, let's do some investigations to actually change some other forms of energy into heat energy.

A Hot Head

Things Needed:
Nail with a head
Sheet of paper

Touch the head of the nail to your lip. It will probably feel quite cool. Now rub the nailhead back and forth across the sheet of paper for about one minute. You should press down quite hard, but not hard enough to rip through the paper. After the minute is up, quickly press the nailhead against your lip. You will find the head is now quite warm. If you continued rubbing for a longer time, you could actually make the nailhead very hot.

While doing this investigation you are getting involved with four different forms of energy. The food you eat provides your body with chemical energy. When rubbing the nail, your chemical energy is changed to mechanical energy. This mechanical energy is then changed to heat energy and, if your rubbing made any noise, you also caused some sound energy.

Perhaps you already know the proper word to describe how your energy changed to heat. The word is *friction*. It occurs when two objects are rubbed against each other.

Long ago man discovered he could actually make a fire from the heat of friction. He did it by rubbing sticks together. His sticks produced heat in the same way your nail did, except his sticks made sawdust. And, by letting the hot sawdust fall on something dry like a leaf, he could fan the spark into a fire. However, you can make a nail hot in a way that can't be done with sticks:

Bending Your Nail

Things Needed:
A nail
A board
A hammer
A pair of pliers

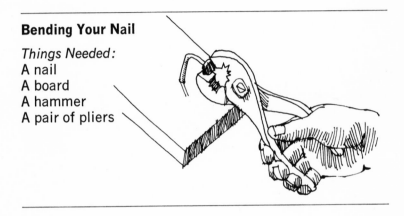

Pound the nail a short way into the board. Pick up the board and touch the head of the nail to your lips; you will notice that it is cool. Grasp the top of the nail with the pliers and bend the nail to the side. Bend it up again, then back. Continue twisting the nail back and forth ten times. Touch the nail to your lips again, and you will discover it has become quite warm.

Tiny molecules which make up the nail attract one another very strongly. This attraction gives the nail a definite shape, and the molecules resist any force which tends to move them. When you bend the nail you must do a good amount of work to change the positions of these molecules. Consequently, the work you are doing is changed to heat energy, and the nail gets warm.

There is still another way to make a nail warm without rubbing or bending it. You do it every time you hammer a nail into a board.

Make Heat With Percussion

Things Needed:
Nail with a head
Hammer
Thick board
   (or several boards
   stacked together)

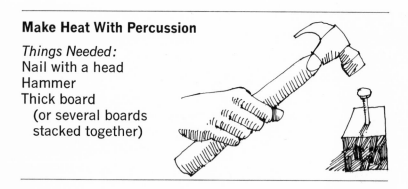

Again, before you start this investigation, touch the head of the nail to your lip to check its temperature. You will find it cool.

Now, simply pound the nail into the board. Pound it hard and keep pounding until it is as far into the board as it will go. As soon as you finish, pick up the board and touch the nailhead to your lip. As before, you will find it warm, perhaps even hot.

Anything that moves has mechanical energy. The hammer demonstrated this mechanical energy as it moved toward the nail. What happened to the energy when the hammer met the nail and stopped? The energy did not disappear, it just changed to another form that you could not see. It changed to heat energy which you only became aware of when you touched the nail to your lip. Also, some of it probably changed to sound energy, as you undoubtedly heard.

HEAT GETS AROUND

All of the forms of energy can move from one place to another. Electricity can travel through wires, sound travels through the tiny molecules of air, mechanical energy moves through gears and levers, and heat can move from room to room. Heat, however, is a difficult thing for a science detective to tail because it can move in any of three very different ways, *conduction*, *convection*, or *radiation*. In fact, the only way to follow it around is to investigate each of the three ways one at a time. Let's begin with an investigation that uses *conduction*.

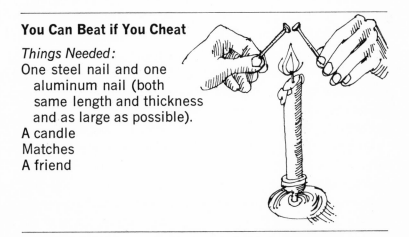

You Can Beat if You Cheat

Things Needed:
One steel nail and one
    aluminum nail (both
    same length and thickness
    and as large as possible).
A candle
Matches
A friend

When you purchase the two nails, be certain that they match in appearance so that your friend will not suspect that one is different from the other.

Give your friend the aluminum nail. Be certain it is the aluminum one! Explain that you are going to have a contest with him. You are both going to hold the point of your nail in your hand. When you say "go" you will both poke the head end of your nail into the candle flame and hold it there. The contest will be to see who has to remove his nail from the flame first.

You will never lose this contest! Your friend will always take his nail away first. He will have to drop it on the table several minutes before you have to remove yours. Of course you have tricked your friend by providing him with an aluminum nail, but it might interest you to know why he had to drop his nail first.

As mentioned earlier, nails are made of molecules. These molecules are sensitive to heat. When you poke a nail into a hot flame the molecules absorb heat energy. They begin vibrating very fast. Of course you cannot see them vibrating because they are much too small to be seen by your eye. Each of the vibrating molecules passes along some of this energy to the ones beside it, causing them to start vibrating. In this way heat energy is passed along from molecule to molecule until all the molecules are vibrating rapidly. The nail is then hot everywhere.

The passing of heat energy from molecule to molecule is called conduction. It is the slowest way of moving heat from place to place.

Some materials conduct heat better than others. You might experiment further by placing different materials in a candle flame. Iron is a poor conductor of heat compared to aluminum. Aluminum molecules pass heat very quickly and easily, as your friend proved nicely. If you have a copper or brass nail you might try these. They, too, conduct heat faster than steel. Investigate to find out if they conduct heat faster than aluminum.

Another interesting way to observe heat conduction is to burn a candle and allow drops of wax to fall on different parts of a nail. Let the wax harden, then place the head end in a candle flame while you hold the point. Watch the globs of wax. As the heat moves up the nail the wax will melt. By watching the wax you should be able to determine how quickly that particular nail conducts heat. You will also be able to decide exactly when you should drop the nail to avoid burning your fingers. If you try this with both the aluminum and the steel nails you will readily understand why your friend lost the contest.

Convection is another way in which heat gets around. It depends on the fact that some molecules such as those in a gas or liquid are not trapped but are free to move from place to place. If these molecules are supplied with heat, they can transfer this energy directly. Let's see how this happens (illustration on next page).

Light a candle. Now hold the point of a nail while you heat its head in the candle flame. Just a few seconds in the flame will suffice. Hold the heated head of the nail beside your nose (be careful not to actually touch it). You will probably feel a little heat. Now, tip your head down and hold the nail *under* your nose, as shown in the illustration. You will feel a great deal of heat. Why should this be?

Roaming Molecules

Things Needed:
A steel nail
A candle
Matches

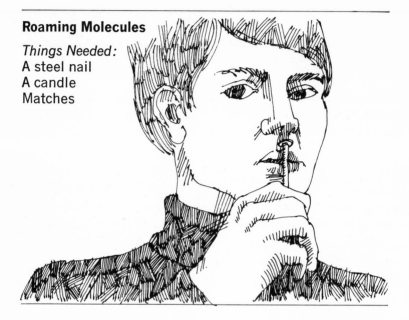

The heated nail is passing its energy to the air next to it. The molecules of air, now heated, start vibrating. As the air molecules are heated they start to rise upward. Heated air is lighter than air that is not heated and the lighter air will rise up through the heavier cool air. This is why you feel more heat from the nail when it is below your nose rather than beside it. The motion of heated air molecules is a perfect example of heat moving by convection.

When your mother heats water in a pan on the stove only the water at the bottom of the pan is heated by the burner, but all of the water gets hot. This happens because the heated water, at the bottom, moves upward through the cooler water at the top. This moves the cooler water down to the bottom where it is heated by the burner. The molecules of water act very much like the molecules of air.

Radiation is the third way heat can get around. It is the most difficult method to understand. Heat can move through empty space where there are no molecules. Radiation uses no molecules.

The sun is almost 100 million miles away from us and there are practically no air molecules in this empty space. Yet the heat from the sun reaches the earth in just about 8 minutes. The sun's heat energy radiates and passes through this space quite easily. This kind of heat is called *infrared* energy. When infrared energy strikes us, or any solid object, we feel the warmth of its rays. Infrared energy is really nothing more than a very special kind of heat.

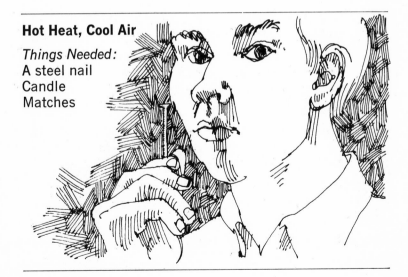

Hot Heat, Cool Air

Things Needed:
A steel nail
Candle
Matches

Light the candle and heat the head end of the nail as in the previous investigation. Hold the nail a short distance away from your cheek so that you can feel the heat radiating from the nailhead. From the previous investigation you might suspect that the heat you feel is reaching your cheek from heated air molecules between the nailhead and your cheek, but this is not the case, and you can prove it! Fan the air with your hand so that any heated air will be driven away. You will still feel the heat from the nail. The heat travels as waves of infrared energy.

While heated air rises, it can also move sideways in a slight breeze. Therefore, the heat you feel beside the nail is actually traveling to you without using the air molecules, similar to the way infrared energy travels from the sun to the earth. Remember, infrared waves are thrown out by any object that contains heat—and this form of heat is called radiation.

So, in review, the nail itself is heated by conduction as the molecules pass along vibrations. The air around the nail is heated and rises by convection. And, infrared heat is radiated outward in all directions and will continue to do so, until it strikes a solid object.

You can think of these three ways heat can move by thinking about putting out a fire with water. Imagine the water as being heat energy and people as being molecules. You could make a long line of people and pass a bucket of water down the line to the fire. That's like conduction. You could also carry a bucket of water to the fire by yourself. That's like convection. You might, however, choose to spray water on the fire from a hose. Nobody touches the water on its way to the fire, and that's like radiation!

LESS HEAT MEANS MORE COLD

If heat is a form of energy, what is cold? Cold is not another form of energy. You are probably just as familiar with cold as you are with heat, yet cold can be more difficult to understand.

The wooden table in your living room probably feels warm. But if you touch the table and then an ice cube, you would definitely say that the table is warmer than the ice cube. You would probably also say that the ice cube was definitely cold. Now suppose you took that same ice cube to the south Pole and the temperature there was 70 degrees Fahrenheit below zero. Do you suppose that same ice cube (at its temperature of 32°F. *above* zero) might feel warm?

Cold is a difficult thing to understand because it does not really mean anything by itself. Objects are cold only by comparison with something that is warmer.

Your body temperature is about 98.6° F. (when you're well). The ice cube is about 32° F. Because the ice cube has a temperature of only 32° F. it has less heat than you have. Therefore, your body is hot and the ice cube is cold.

The sun's temperature is about 9,900° F. Compared to you the sun is very hot and you are very . . . very . . . cold.

Cold is just a word used to compare the amounts of heat different objects contain. A cold object actually contains heat energy but if it contains less than another object we say it is colder.

So, the less heat an object contains, the colder it is. If you are still confused by this definition of cold, here is an investigation to try:

Putting Your Heads Together

Things Needed:
Two nails, with heads
A candle
Matches

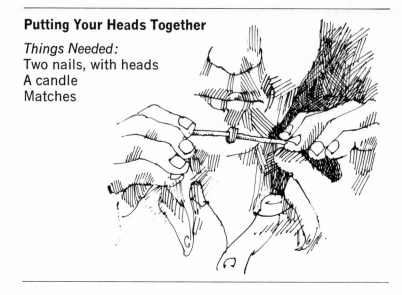

Light the candle. Hold the points of the nails, one in each hand. Touch the heads of both nails to your cheek to check their temperatures. Again, they will feel somewhat cool. Now hold the head of one of the nails in the candle flame for about 30 seconds. Remove the nail from the flame and quickly press the heads of the two nails together. Hold them tightly together while you count to 25. Touch the head of the nail that was *not* held in the flame to your cheek. You will notice that it now feels warm.

Earlier in this chapter you learned that heat energy causes molecules to vibrate faster. When you press the heads of the two nails together, some of the vibrating molecules in the heated nail contact those in the un-heated nail and cause them to vibrate also. In this way

Keeping A Cool Head

Things Needed:
Two nails
An ice cube

some of the heat energy is conducted to the second nail. Then, when you held this nail to your cheek you were able to detect these vibrations as heat.

Now let's try the investigation above:

Repeat the previous investigation, but use an ice cube instead of a candle. Touch both nails to your cheek and check their temperature. Hold one against the ice cube for 30 seconds then press its head against the head of the other nail and count to 25. Touch the head of the nail that did not touch the ice cube to your cheek. You will find that it now feels much colder than it did originally.

Now, before you read much further, please try to explain what happened in this investigation. Did you really pass the cold from one nail to the other? Can you really pass cold along in the same way you passed heat in the previous investigation? Be careful with your thinking before you answer. Many people make a mistake in thinking about this problem.

You read the answer to this mystery earlier. All cold objects actually contain some heat energy. You must remember that heat is a form of energy and energy can be passed from one object to another. The coldest nail in this investigation is simply the nail which contains the least amount of heat energy.

What really happens? When you press the first nail against the ice cube, the nail has more heat energy than the cube. The heat from the nail moves into the ice cube and starts the water molecules vibrating.

There is a science law which says that heat will always move from an area of heat into an area of less heat. In fact, if a hot and a cold object are together, the heat will move from the hotter to the colder until both objects are the same temperature.

This simply means that the ice cube does not pass its cold to the nail, rather the nail passes its heat to the ice cube. The nail later feels cold because it contains less heat energy.

Next, when you press the heads of the two nails together, the cold nail does not pass its cold to the warm one. The warm one must pass its heat to the cold one according to the science law. In passing its heat, this nail then becomes cooler.

So, surprisingly, the answer to the original question is No, cold cannot be passed along like heat. Only heat can be transferred from object to object but it is easy to falsely assume that the cold has been transferred.

Do you truly understand this puzzle? Here are three examples which are true. If you understand them, then you know what cold is:

• Put your finger on an ice cube. The cube feels cold because your finger is warmer. The cube is not passing cold to your finger, it is taking heat from your finger, so your finger *feels* cold.

• Relative to the sun our bodies are very cold. The heat from the sun passes to our cold bodies in summer, so we feel warm. In the winter we feel cold because we are passing our heat to the air which, relative to us, is cold.

• Scientists can make air a liquid by compressing it and making it very cold. Liquid air is about 320° F. below zero. If you placed a pan filled with liquid air on a block of ice (at 32° F.) the liquid air would boil just like water on a hot stove. The liquid air boils because the ice is hot.

A COOL PUZZLE

You have already learned that different materials conduct heat at different speeds. Because of this, the problem of deciding what is hot and what is cold becomes even more difficult. One way a scientist never uses to actually determine temperature is by touch, or feel.

Special nerves in your skin are able to detect temperatures, and they generally do a pretty good job for you, but these temperature sensors are easily fooled. Here is an old party trick you can do, using nails, that will prove this:

How to Make a Friend Burn

Things Needed:
Two matching nails
A candle
Matches
A really good friend

Read this trick carefully to be certain you understand it before you actually try it on your friend.

Tell your friend that you wish to try an investigation with a nail. Hold the point of one nail in your hand and start heating the head in the candle flame. Be sure your friend watches you do this. The other nail should be hidden in your pocket, and your friend should not know that you have this extra nail. When the first nail has heated for a short while, ask your friend to shut his eyes. He will probably be somewhat reluctant to do this because he is afraid you are going to burn him. You may assure him (quite honestly) that you will not. When his eyes are shut, lay the heated nail aside, and remove the cold nail from your pocket. Touch the *cold* nail on his hand and say "Now let's see what happens when this nail touches your hand." I'll bet your friend jumps, grabs his hand and gives out with a noisy yell. Naturally, you should immediately explain to him exactly what you did, and assure him that he is not burned (particularly if you want to keep him as a friend).

The interesting part of this investigation will come when you discuss what your friend felt. He will probably tell you that the cold nail felt hot. Part of the hot feeling was in his mind, and part in the temperature sensors in his skin. You have proved, in a rather dramatic fashion, that skin is actually not a very good temperature indicator and cannot really be trusted for deciding what is hot or what is cold.

Naturally, you cannot fool your friend again with this investigation, nor can you ever fool yourself with it. However, if you would like to see, for yourself how difficult it is to judge temperature, here is a cool puzzle for you to try:

What's Hot, What's Cold

Things Needed:
A nail
Anything else in your room

Begin by touching the nail. Is it hot or cold? Touch the floor. Is it hot or cold? Touch the walls, the furniture, a drinking glass—touch everything you can in the room (except please don't touch a radiator, any part of the heating system or a lighted bulb). What objects are hot, which are cold? Which is the warmest? Which is the coldest?

Please touch lots of objects in your room before you read what follows. Once you have decided which objects are hot and which are cold, you will be very surprised with the next paragraph.

If you have a thermometer in the house put it in your room. If not, let's imagine that this thermometer reads 70° F. If you do have one, it also will probably read about 70° F. You can put this same thermometer near the floor and it will read around 70 degrees F. You can put it on the table and it will read around 70° F. You can lay it on the nail and it will still read around 70° F. Almost anywhere you put the thermometer the reading will be around 70° F.

Except for the heating system, and the hot light bulb, every object in your room is at room temperature. None of the objects can make any heat of their own, so they can only absorb the heat that is available. Consequently they will all become the same temperature.

Why do some feel cooler than others? Because you are warm blooded and, you are a different temperature. Being warm blooded means that some of the energy you obtain from eating food is changed into heat. When you touch an object that is not as warm as you are, some of your heat is immediately conducted to that object. If it is an object like metal, which conducts heat very easily, it will absorb your heat quickly and you will say it feels "cool". However, if it is made of a material that does not conduct heat well, such as wood, heat is not conducted from your hand so the object does not feel cold.

We use our own body temperature, or 98.6° F., to judge the temperatures of other objects. Objects having less heat than ours, will absorb our heat. Depending on how well they are able to do this, they may feel cooler or warmer than they actually are.

4
Magnetic Nails

Magnets! The very word usually interests a science investigator. Certainly everybody who is familiar with magnets is curious about their mysterious attractions and repulsions. And, because we can't see, taste, touch or smell magnetism, the word becomes a marvelous mystery.

Where does magnetism come from? What are its properties? Just exactly what is magnetism? These are the kinds of questions you probably asked after experimenting with a toy magnet.

What do you require to study magnetism? If you wished to really impress your friends you might experiment with a giant magnet, on a crane, that is powerful enough to lift an automobile. But, if you would prefer to investigate alone, in your room, you will find the following investigations much more practical. Because they use only tiny magnets, and few materials, they might be called micro-investigations. To do them all will require only some nails, a file, a flashlight battery, a short piece of wire, and a sheet of aluminum foil. If you can locate these items you will be ready to begin your study of a fascinating science subject.

MAKING MAGNETS

Magnets are nothing new. Very ancient people were well acquainted with their strange properties. Certain rocks, ores of iron, were found in various parts of the world. These rocks were called *lodestones*. People soon discovered that lodestones would attract bits of iron. This attraction was so mysterious to these people that many superstitions developed around the rock. One such superstition was the belief that there were enormous mountains of lodestones rising out of the sea. Many sailors would not venture too far from land, believing that if they came too near one of these mountains, the iron nails might be drawn out of their wooden ships! Naturally, this belief was a foolish and unfounded one, but the fascination people still have for the mysterious lodestone has continued through many centuries.

What is magnetism? Surprisingly, no one really knows. Scientists know how to perform many jobs with magnetism but, they do not know what magnetism *is*.

I once knew a boy who told me that he liked science but "did not want to be a scientist because other scientists had already learned everything, and there was nothing new to be discovered". It surprised him to learn how many things scientists do *not* know: What gravity is, what electricity is, where the earth came from, what makes lightning, and many many more.

Perhaps your job in the future might be to answer the question that faces us now . . . what is magnetism?

Although they don't know what magnetism is, scientists do know a great deal about it. They know, for example, how to make a bar of steel magnetic. The two common ways they perform this science-magic trick is by *induction* or by using *electricity*. You should investigate both methods.

First though, you will require some iron filings. You will be using them for many of the investigations that follow.

Make Some Iron Filings

Things Needed:
A nail
Hammer
Piece of board
Sheet of white paper
A file

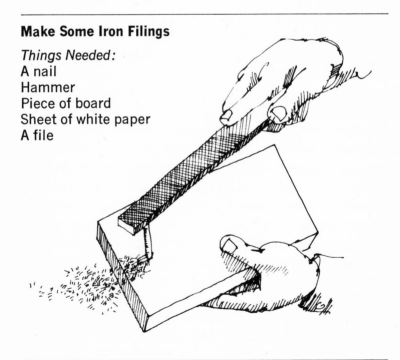

Pound the nail a short way into the board. Hold the board, as shown in the illustration, tipped up on the sheet of paper. File the top of the nail, making sure that you catch all the tiny filings from the nail on the sheet of paper. Save the filings. Continue filing until you have enough filings to cover your fingernail. Be sure to save the nail and board just in case you need more as your investigations continue.

Now, choose a good size steel nail that you would like to magnetize. Touch the tip of the nail to the iron filings; you will notice that it does not attract them. The nail contains the properties to become a magnet, and your job now is to prove it.

Scientists have long known that magnetism and electricity go together. Wherever there is electricity, there will be magnetism. A good place to begin your study of magnetism is by learning a little about electricity.

Electricity is the motion of *electrons*. Electrons are the tiny bundles of energy that swirl about atoms. Because the electrons are moving around the atom, every atom contains some electricity. A nail, for example, is filled with electricity made by the motions of the electrons around the iron atoms. Because the electricity is in such tiny quantities, you are not aware of its presence. But because electricity is there, you would also find something else around the iron atoms, you would find magnetism.

Perhaps you should prove to yourself that magnetism can be found wherever there is electricity. Here's an investigation that will convince you:

Electricity Makes Magnetism

Things Needed:
Piece of insulated wire
   about two feet long
A flashlight battery
Iron filings
   made previously

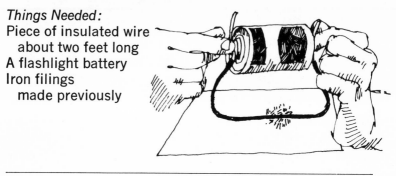

Scrape one-half inch of insulation off each end of the wire. Touch the center of the wire to the filings to convince yourself that there is no magnetism in the wire. Now, touch the ends of the wire to the battery, as shown, and allow a current to flow through the wire. While the electricity is flowing, again bring the center of the wire near the filings. The filings will jump and stick to the wire. Disconnect the wire from the battery, give the wire a little shake, and the filings fall off. The magnetism is only there while there is electricity flowing through the wire!

Note: Do not leave the wire connected to the battery for more than a few seconds. This causes a short circuit, and will quickly run the battery down.

The same magnetism that surrounded your wire, when electricity flowed, also surrounds the atoms in a nail. Electricity, in the form of electrons, is always present in the nail, so there is always magnetism around the atoms. Why then isn't the nail a magnet?

Think of the atoms as being pointed in different directions and at different angles. Their magnetism also points in different ways. Without any regular order, one tiny atom magnet is actually fighting against another. Unless a great many of the atom magnets can line up and work together, we are not aware of any magnetism. The easiest way to make them get in line is by pushing against the electrons around the atoms. You can't see electrons, but you can still give them a shove.

Pushing Electrons Around

Things Needed:
Battery, wire, and iron filings
   from last investigation
Large steel nail

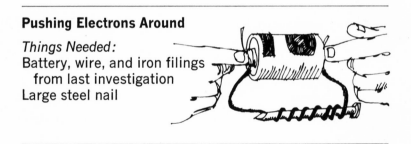

Wrap the wire around the nail as many times as you can. Touch the ends of the wire to the battery as you did in the previous investigation. Leave the wire connected to the battery while you count to ten. Disconnect the wire and touch the nail to the filings. If you have succeeded, the nail will attract the filings. You may now unwind the wire and save your magnetized nail for further investigations. If you did not succeed, try again, counting a bit more slowly, (or trying another battery) until you do magnetize the nail.

What happened here? Before you started, groups of atom-magnets were scattered throughout the nail. These groups were lined in the same direction (scientists call these groups *domains*) and the atom-magnets in the domains were working together. Of course, other domains were pointing in other directions and they cancelled one another out. When you applied electricity to the wire, the wire was surrounded with magnetism. This magnetism pushed against the magnetism in the domains. The domains shifted their position and thousands of domains, which had been fighting one another, lined up in the same direction and started working together. For the first time there was enough magnetism in the nail, facing the same way, to prove its existence.

Remember, you have not "made" any magnetism. The magnetism was in the nail all along. You have simply organized it and made it do what you wanted.

There is still another way you can push magnetism around and make the domains get into line. This is called magnetic *induction*.

Rub Yourself A Magnet

Things Needed:
Magnetized nail from
  last investigation
A similar nail that is
  not magnetized
Iron filings

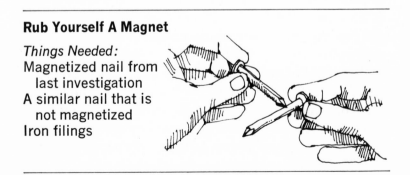

Induction can change a bar of iron into a magnet by simply bringing it near an object that is already magnetized. Hold the head of the unmagnetized nail in one hand, and the head of the magnetized one in the other. Then rub the point of the magnetized nail down the length of the other. In this way, you should be able to swing some of the magnetic domains into line in the other nail. Before you try this, however, here are a few tips you should follow if you want to be successful: Be sure to rub in only one direction. Touch the point of the magnetized nail to the head end of the other. Slide the tip down the nail and off the end. Lift the magnetized nail away, and again touch its tip to the head end of the non-magnetized nail and repeat the procedure. Continue this for 50 rubs. Always rub the non-magnetized nail from head to point and, once you've started, never rub it the opposite way. Also, make certain to rotate the non-magnetized nail as you rub it. Be sure to rub all sides of it with the magnetized one.

After rubbing 50 times, try picking up some iron filings with the nail you have been rubbing. It will attract some of the filings, and you have indeed helped it become magnetic, but it will not be as strong as the magnet you made with electricity.

MAKING MAGNETS STRONGER

You may be a little disappointed with the magnets you are able to make at home because they are usually weak compared to store-purchased ones. Remember though, your magnets are made in the same manner and they work exactly the same way. If you understand your own micro-investigations you will also understand a great deal about all magnets.

The most powerful magnets are made by placing a bar of metal inside a coil of wire and passing an electric current through the wire. Why are they stronger than yours? To make a powerful magnet requires the use of enormous currents of electricity, far stronger than anything you could provide at home. The more electricity used, the more magnetism surrounding the wire. The more magnetism, the greater the number of domains brought into line—so the final magnet is stronger. Also, special *alloys* (mixtures) of metals are used instead of plain steel. These alloys provide domains that tend to remain in line better than steel. One such metal is an alloy made of aluminum, nickel, cobalt, copper, and iron. It is called *Alnico*, and Alnico magnets are among the strongest available.

If you use nails, your magnets will be of one metal, iron. You can, however, use a greater current of electricity to magnetize them. This should result in a stronger magnet.

A Stronger Magnet

Things Needed:
A large nail
Wire from last investigation
Two flashlight batteries

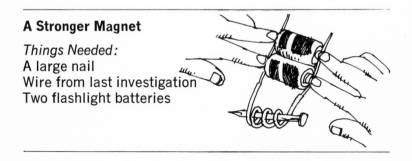

Imagine the batteries as buckets filled with water. When you attach the wire to the battery a current of electricity runs through the wire, just like a current of water would flow if you tipped over a bucket of water. Two batteries contain more electricity, just as two buckets contain more water. By fastening the wire to the batteries in the proper way you can dump the electricity out of both at the same time and provide a stronger current.

One end of the bared end of the wire must touch *both* batteries at the top end. The other end of the wire must touch the bottoms of *both* batteries. Be sure the insulation, on the wire, has been removed where it touches each battery. This is called a *parallel* arrangement of wire and batteries. This allows twice the current to flow.

Use the batteries and the wire in this parallel arrangement and wrap the center of the wire around the nail. You will proceed to magnetize the nail in the same manner as you did, using only one battery, earlier in the chapter. Your new magnet should attract more filings than your earlier one.

Here is another way to make a stronger magnet.

One Magnet From Two

Things Needed:
Two magnetized nails
Iron filings

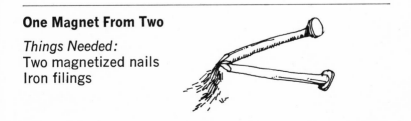

Press the two magnetic nails together, tips touching, and bring them near the filings. Each nail is able to pick up just as many filings as it did before, so two together should be able to pick up twice as much. This should suggest another way of making a stronger magnet.

Suppose you had used a larger nail in your investigations. There would be more domains available to be lined up and, perhaps you could think of the large nail as being made of two smaller ones. Why not try magnetizing a spike, or similar large nail, and see if this doesn't result in a much stronger magnet?

Of course, to magnetize a larger nail you might wish to use still more current. Perhaps you might use three or four (or more) flashlight batteries. Arrange them all together in a parallel wiring and you may be surprised at the strength of a nail magnet you can create.

THE END OF THE MAGNET

Our earth is a giant magnet. No one is really sure, but the earth probably obtains its magnetism from electricity made by molten iron and nickel sliding around inside it.

The earth's magnetism is centered at an area in northeastern Canada, by the North Pole (with another center by the South Pole). The needle of a compass points, not to the North Pole right at the top of the earth, but to the *magnetic north pole* which is off to the side. The ends of a magnet are named after the North and South Poles of the earth. The actual words, north and south, are just handy names a scientist gave the different ends long ago so he could talk to people about them; and when referring to a magnet, they mean just this, its two ends. They can help each other, or they can fight against themselves. You can do an investigation to show this:

North Versus South

Things Needed:
Two magnetized nails
Iron filings

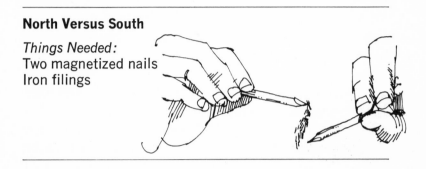

Pick up some iron filings on the point of one of the nails. Try to pick up enough so that a long "beard" hangs down from the tip. Now, bring the point of the other nail toward the filings. One of two things will happen, the iron filings may move away from the tip of the approaching nail, or they may be attracted and jump toward it. Next, reverse the second nail and bring its head end toward the filings hanging from the first nail. Whatever happened the first time will not happen this time, but the opposite thing will. One end of the nail attracts the filings, the other end pushes them away. There is an obvious difference between the head and point of the nail, one end is the north pole, and one is the south.

To explain what you have observed you should learn two very simple rules regarding magnets: *unlike poles attract,* and, *like poles repel.* The rules are simple to remember, and simple to understand. If you have two magnets, then you must have four poles, two north poles and two south poles. The two south poles would be the like poles (they are the same, alike) so they will repel, or push away, from each other. Two north poles are also like poles and they too repel each other. If a north and a south pole come together, however, they obey the other rule, and the two unlike poles will attract each other.

The iron filings were on one pole of your nail magnet. If you brought the point of the other nail near them, and it was a like pole, the filings were pushed away. If it was an unlike pole it attracted the filings. Reversing the second nail simply proved both of the rules.

If you understand these basic rules you are now ready to move along and learn still another fact about the poles of a magnet. Here's an investigation you can do, and it will work well for you. Yet it is not really true at all. Think about that as you proceed.

Where There's No Pole

Things Needed:
Magnetized nail
Iron filings

Is there a place on a magnet that has neither a north nor a south pole? There certainly appears to be. If you spend a moment with your nail magnet and the filings perhaps you can find it before you read any further. Perhaps you can even think of where it might be without actually experimenting.

Attract some filings to one end of the nail magnet. Now, attract some to the other. You can do that without any trouble. Next, let's see you attract some filings to the middle of the nail. You'll discover that you can't do that!

If one end of the magnet is north, and the opposite end is south, what pole would you find in the middle? You are probably tempted to say it is a spot with "no poles at all", but, if you say that, you are absolutely wrong. There is no place on a magnet without poles because of the magnetic domains.

Make A Breaking Magnet

Things Needed:
Two nails
Sticky tape
Flashlight battery
Piece of insulated wire

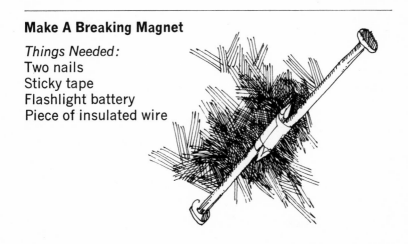

Overlap the two points of the nails as shown in the illustration. Wrap them with sticky tape so they are fastened together. Be sure that the metal of each nail touches the other when they are taped together. The two nails are now like a single bar of iron, and the bar can be magnetized just like a single nail.

Wrap the wire around the double nail bar and, using the battery, magnetize it as you did in previous investigations. When you're through, touch each end to the filings and make sure it is a good magnet.

Try to attract some filings to the center of this magnet, where the sticky tape is. You will discover that, like a single magnet, the middle does not attract the filings. Once you've proven this, you are ready to prove that the middle of a magnet is magnetic.

Remove the tape and separate the nails. Try to attract filings to both ends of both nails. You'll find that all four ends now attract filings. It would seem that poles have suddenly appeared on the ends of the nails that were originally the middle of the long magnet. Where did these poles come from?

Wouldn't you guess that they were really there all along? Where does the north pole, or the south pole, stop on a magnet? Certainly it is not just on the ends, because even the sides toward the ends will attract filings. Maybe they extend down to the middle of the magnet and, in the middle they simply cancel each other out. Perhaps the best way of thinking about this is with a drawing of an imaginary magnet. Each of the tiny domains in the magnet act like tiny magnets themselves. Once they are lined up, the north poles point one way, making that end north, and the south poles point the other. If you study the drawing you can understand why the poles appear on the ends, but do not have much affect in the middle. You can also understand why, if you "broke" the magnet in half you'd produce two magnets, both with north and south poles.

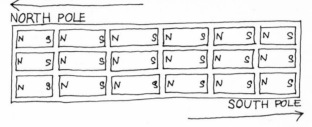

Finally, here's a puzzle to prove whether or not you really do understand this idea of magnetism: Suppose you had two nails, one a strong magnet and one not magnetized at all.   How could you, without using anything but those two nails (no filings or the like) decide which was the magnetized one?   Remember that the magnetic nail would stick to the non-magnetic one and vice versa, so you could not decide simply by sticking their ends together. This is a nice "no poles at all" puzzle to leave with you.

NORTH OR SOUTH?

You never have to identify which end of a magnet is north and which is south, but by now you are probably wondering how it could be done. All you need is one magnet, with it's north pole marked, and you can easily identify the poles of all your other magnets, but how do you find the north end of the first magnet?

Why does a compass needle mysteriously point toward the direction of the magnetic north pole? The needle is actually a miniature magnet and, like your nail magnets, it has a north and a south pole end. The end of the compass needle that is attracted toward the north pole of the earth is called the north seeking pole, or *north pole* for short.

So it is possible to make a magnet point out it's own north pole. All you have to do is make your magnet into a compass.

North To The Pole

Things Needed:
Magnetized nail
Piece of aluminum
    foil, 4″ square
Glass or plastic
    bowl of water

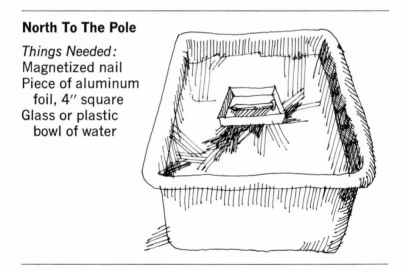

Fold the sides of the aluminum foil up to create a square boat that will float on the bowl of water. Lay the magnetized nail carefully in the floating boat and move the boat to the middle of the bowl. Do not touch anything for ten minutes. The boat will slowly swing around and one end of the nail will point northward.

You must know what direction north is. If you don't, go outside early in the morning and face toward the rising sun. Because the sun rises in the east, north will be to your left.

Whichever end of your nail points northward will be the north pole of your nail magnet. If you wish to double check your results, turn the boat around and leave it alone for another ten minutes. You should find that the same end always swings around and points north.

If you are not successful with this investigation you might check the following: Be sure the bowl you are using is made of glass or plastic. If you try it in a metal pan or sink, the magnet could be attracted by the metal. Also be sure there are no large metal objects near the bowl that might also attract the magnet. If there are no metal objects close by, your boat should stay right in the middle of the water (be sure it is perfectly still). If the boat drifts to one side, move your bowl to a new location.

Once you know which end is north, dab a bit of paint, or your mother's nail polish, on it to mark it permanently. You can use this one nail to identify the north pole of any other magnet you may have. To use it, just attract some iron filings to the end you've marked north. Bring the end of another magnet toward the filings. If the filings are repelled then it is the north pole of your unknown magnet (like poles repel), if it attracts the filings it's opposite end must be the north pole (unlike poles attract).

WHAT MAKES A MAGNET?

Here's a statement to think about: You can make a magnet, but you cannot be a magnet. You've already made a magnet, so the first part is true, but if you wrapped a wire around your finger, and attached it to a battery, do you suppose it would attract bits of iron? You can't be a magnet. If you don't believe it, try it!

Everything is made of atoms surrounded by whirling electrons. We've already learned that magnetism surrounds whirling electrons. Everything, then, contains some magnetism, but the magnetic domains of some materials are more easily lined up than in others. Iron filings are attracted to a magnet because, as a magnet approaches them, domains within the filings swing around so that they become tiny magnets themselves. The tiny magnets within the filings attract the magnet, and the magnet attracts them. If the magnetic domains within the filings were not free to swing around, the attraction would not occur.

Discover Magnetic Things

Things Needed:
Magnetized nail
Objects around your house

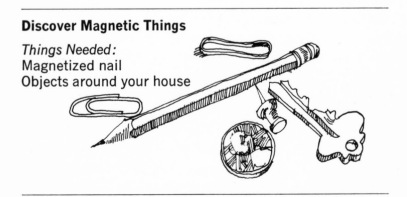

Be sure to use your strongest nail magnet for this investigation. How many objects around your home will the magnet attract? I'll bet you will find more objects that aren't attracted than are. Try this list: A staple, a bit of steel wool, a tiny scrap of paper, a piece from a plastic bag, a penny, a bit of cloth, a paper clip, a piece of pencil lead, a tiny pebble, an elastic band, and a blade of grass. Actually, your magnet will only attract three items on the list. Try to discover others to add to it.

The best magnetic materials are called *ferrous* materials. Ferrous materials are all, or contain, iron. Iron is a strongly magnetic material because of the nature of its magnetic domains. Other strongly magnetic materials, besides iron, are nickel, cobalt, and an unusual substance called gadolinium. All other materials contain magnetic atoms, but they are not strongly magnetic. Such things are called weakly magnetic materials. Glass, wood and cloth are common examples of weakly magnetic materials.

Weakly magnetic substances, even if they're made of metal, do not make good magnets. You can prove-it-yourself with some different kinds of nails.

Bet You Can't Make a Magnet

Things Needed:
Wire and battery from
    previous investigations
Nails made of aluminum,
    brass, or copper

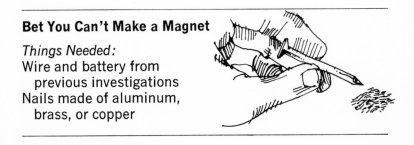

This is an investigation that will not work for you! You might, however, want to try it just to prove that very thing. Wrap the wire around one of these non-ferrous nails, attach the wire to the battery, and attempt to magnetize the nail as you did before. When you're done, you will find the nails do not attract iron filings.

HOW TO LEARN MORE

Magnetism is a vast subject. This chapter has barely served to introduce you to some of its basic ideas. Many scientists have spent their entire lives studying magnetism, and they still want to learn more. If you would like to continue learning and studying magnetism, here are some ideas for investigations:

- In the process of making a nail magnet you wrapped a wire around a nail and connected it to a battery. While the wire is attached to the battery the nail is an *electro-magnet*. If you bring the nail (while the electricity is flowing) toward the filings you will find the magnet nail much stronger than when the wire is disconnected.

- Which is best for attracting filings: the sharp point of the nail or the flat head?

- Pound one of your nail magnets into a board, then pull it out with a hammer. You will find that it is no longer magnetic. The pounding has destroyed its magnetism. Can you explain why?

- Magnetize a paper clip by wrapping it with wire and attaching the wire to a battery, as you have done with the nails. Now straighten the clip out. Sprinkle filings along its length and you will find several places where filings are attracted. This is called a *multi-polar* magnet.

- Here's a difficult statement to explain: If the earth is a giant magnet, and the north pole of a compass needle points toward the north pole of the earth, then the magnetic north pole of the earth must really be the south pole of the giant earth magnet!

This statement is true, but it's a tricky one to leave with you. Perhaps it will remind you again that there are many ideas remaining to be thought about regarding magnetism. Everyone has played with magnets, but how well do we really understand them?

5

Math Magic With Nails

Mathematics is a subject that usually deals with numbers, multiplications, divisions, subtractions, additions, and the like. Maybe you will be surprised to find a chapter dealing with the mathematics of nails.

Actually, mathematics is not just a subject you study in school. It is an important part of your everyday life. If you drive one nail through two boards you are working with mathematics. If you use ten nails to make a box, again mathematics is there. If the hardware store man sells you a pound of nails, a lot of mathematics is involved. There are lots of times when nails and mathematics work together. You will be using all of the ideas suggested above, plus a few others, as you do the investigations that follow.

A HAMMER AND NAIL PUZZLE

If you have a friend who feels he is particularly clever with mathematics, here is a puzzle that may stump him:

Three Boards and Six Nails

Things Needed:
Three small pieces of thin wood
Six nails
A hammer

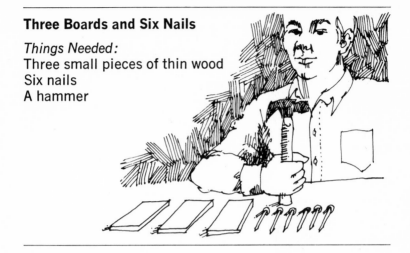

Give your friend the following challenge: He is to pound the six nails into the three pieces of wood. He is free to drive as many nails as he wishes in each board, but he must put at least one nail in every piece. Now, this is the real problem. When he is done, every board must have an *odd* number of nails going into it (that is, each piece must have either one, three, or five nails going into it). No piece should have an even number (two, four, or six).

Think about this a moment and you may conclude that it is impossible. Perhaps you might guess that there is a trick to it, and indeed there is, but it is an honest trick.

Have you ever wondered how a carpenter can join together an even number of boards (two) with an odd number of nails (one)? Of course you haven't because the solution is too obvious. He just puts one board on

top of the other and drives the nail through both of
them. This, (because it is actually a clue,) will help you
to solve the puzzle of the three boards and six nails.

Actually, there are many ways this problem can be
solved successfully. Here's one: Drive two nails a short
way into two boards toward their ends. Lay one of the
boards on top of the other, and pound the fifth nail
through them both. The final nail goes into the third
piece of board. Now every board will contain an odd
number of nails. That is, one nail in one board, and
three nails in each of the others. Like this:

This is just one combination that would result in a
satisfactory solution to the problem. Perhaps you can
think of another that would also do the job. Here, for
example, are two other ways to do it:

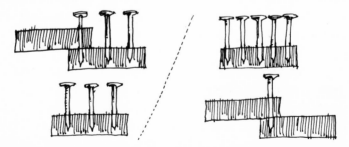

Perhaps you can think of another way to solve the
problem. So far you have been using two pieces of
wood together, and one piece separately. There are

many ways this will work; but you could do it by stacking all three pieces together and pushing them in and out. Use these drawings if you need to.

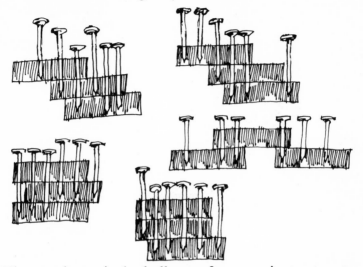

The mathematical challenge for you is to try to discover as many different ways to solve the problem as possible. Once you understand the solution to the problem the challenge is to try to find as many different combinations that you can. You have seen quite a few solutions, but there are many more.

THE GAME OF FIFTEEN

There is a very ancient game, called *Nim*, or *Fifteen*, which has been played by many people for many years. To begin, find fifteen of any object. You certainly can find fifteen nails, and if you've never played the game, here's how:

Pick Up Nails

Things Needed:
Fifteen nails of any size
A friend to play with

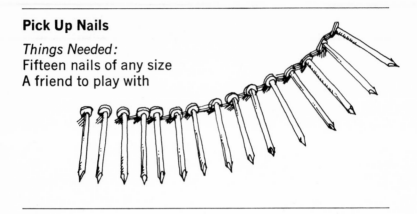

Lay the nails out in a row, as shown in the illustration. Explain the rules of the game to your friend: each of you will take turns picking up nails. Each player will pick up three, or fewer, nails on each turn. The player must always pick up at least one nail, but never more than three. The object of the game is to leave just one single nail for your opponent to pick up. The person who picks up the last nail is the *loser*. Either of you may start.

Before reading on you might like to try the game several times. There is a mathematical secret which will allow you to win every time, but the game is really more fun when you do not know the secret.

The secret for winning is very simple, all it requires is a bit of studying and figuring on your part. For this reason, I will simply tell you how to win and leave the question of why you win for you to ponder.

First, remember the numbers 13, 9, 5, and 1. These are the key numbers to the game of fifteen. No matter who starts, when it is your turn leave one of these numbers of nails on the table. You will find that you can do this without any problems. If you abide by this routine you cannot help but win.

There is one exception to this rule. Suppose your friend also read this book, and decided that he would leave you with 13, 9, 5, and 1 nails. Who would win then? If both of you knew this winning formula, the person who picks up nails the first time will always win. The reason for this will be very clear if you will lay down 15 nails and try to play yourself using the formula.

After you play, you will understand why this formula is always a winning one. What is the mathematical reason for its success? Now, just for fun, see if you can work out a formula you might use if you wanted to lose every time.

MATHEMATICS WITHOUT NUMBERS

It is difficult to think about mathematics without thinking about numbers. But consider doing mathematical investigations in which you never use one single number. If the idea of number-less mathematics appeals to you, you might consider the mathematics of *geometry*.

Geometry is the study of points, lines, objects and the like. It studies space that things occupy, and the shape of them. Numbers are often useful in studying geometry, but you can do some simple geometric investigations without them.

Surprisingly, to those readers who don't particularly care for mathematics, many games and toys use geometry. There's a very good chance that some game you've enjoyed involved some geometry; you were actually studying the subject without even knowing it. The following investigations and puzzles will seem like such games to you. Once you become involved in solving the problems you will not realize that you are actually studying the relationships between straight lines, squares and triangles—all shapes basic to geometry.

Introducing A Puzzle For Squares

Things Needed:
Twelve nails

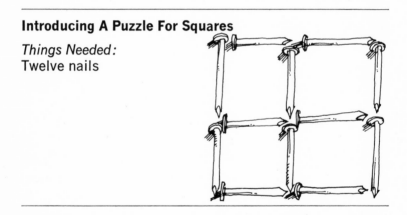

To begin, lay the twelve nails on a table so that they make the design shown in the illustration. Be certain that it matches exactly.

Now, remove just four nails from that design so that the eight nails remaining on the table form just one single square. This is the first puzzle, and just to start you off it is an easy one. All you have to do is take away the four nails in the center, like this:

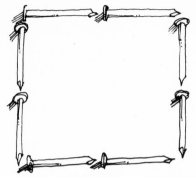

Now, let's get a little more complicated. Remove just two nails, from the original figure, to leave *two* squares remaining on the table. Here's how this would be done:

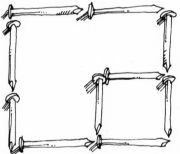

Part of the large square forms two sides of the second, smaller, one. This is a geometric trick you may find useful later on.

Now that you have the idea you can add more nails, make the puzzle more complicated, and perhaps invent further puzzles yourself. Challenge your friends with them; you will be surprised how fascinating these tricks will become.

Nine Squares For More Fun

Things Needed:
Twenty-four nails

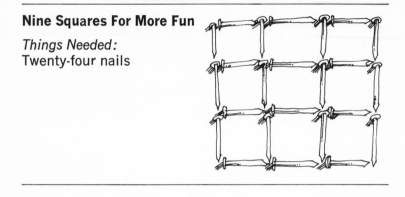

Arrange the nails as shown in the illustration. You will use all 24 nails to make nine squares if you have arranged them properly. Now, rather than showing you how to solve the puzzles, let me list a few possible ones. Try to solve these first, then try to invent others of your own:

1. Remove 12 nails and leave 1 square
2. Remove 16 nails and leave 1 square
3. Remove 20 nails and leave 1 square
   (Those were the easy ones)
4. Remove 8 nails and leave 2 squares
5. Remove 8 nails and leave 3 squares
6. Remove 8 nails and leave 4 squares
7. Remove 4 nails and leave 5 squares
8. Remove 6 nails and leave 3 squares

All of these problems can be solved, none are impossible. Now, before leaving you to create your own "Geome-tricks", I would like to remind you that squares are not the only shapes you can create with nails. Perhaps you might wish to create puzzles with another shape. For example, here is a pattern you can make:

Now Try Triangles

Things Needed:
Sixteen nails

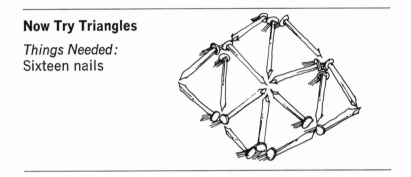

Arrange the nails to make the pattern shown in the illustration. Just to get you started, can you:

1. Remove 5 nails and leave 4 small triangles
2. Remove 4 nails and leave 5 small triangles
3. Remove 3 nails and leave 6 small triangles
4. Remove 2 nails and leave 7 small triangles

Of course, you will find these are all easy ones. For a toughy, try removing six nails leaving you with two triangles. Or, even harder, how about removing four nails and leaving four triangles?

As you create other designs, and other puzzles with them, you will appreciate the fun you can have with some simple geometry.

Just for fun, after you've shown some of these geometric puzzles to a friend, hand him 11 nails and ask him to lay them down on the table in such a way that the 11 nails will become one. When he gives up, show him . . .

WHISPERING NAILS

The science magician (that's you) explains to a friend that he can actually "talk to nails" and that they can "talk back to him". When your friend stops laughing you will then proceed to convince him that you were actually telling the truth (illustration on next page).

You do it in the following way: While his back is turned, instruct your friend to reach into a dish full of nails and remove any number he likes. You should turn your back also so you can't see how many nails he has removed. The number should be more than ten.

You ask your friend to count the number of nails he has removed; then tell him to return some of the nails to the dish in the following way. He determines how many to return by adding the two parts of the number he originally selected. For example, if he picked up 15 nails he would return six nails to the dish (because one and five equals six). If he chose 18 nails he would return nine nails (one and eight equals nine), and so forth.

Of the nails he has left in his hand he can keep any number he likes, but he should hide them so you can't see them. If he has any left over he should hand them to you. Once he has done this you turn and face him for the first time since you've started.

You look at the nails he has given you. You ask the nails "How many nails did he hide?" You hold the nails close to your ear as though they were whispering to you—and then you tell your friend exactly how many nails he is hiding!

Nails In Your Hand

Things Needed:
Nineteen nails
A small dish to hold them
A friend

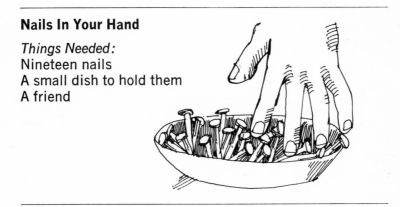

As you know, there is no magic to this trick, no sleight of hand, no peeking, and you will not require any practice to do it correctly every time. This is a mathematical trick, and this one uses numbers.

All you must do is have your friend do the exact steps outlined above. Once you receive the "whispering nails" from your friend you need only perform a simple bit of arithmetic to determine how many nails he is hiding.

Maybe you already discovered the magic to this mathematical mystery. First, by having only 19 nails in the dish and asking him to pick up a number more than 10, you have forced him to select a number between 11 and 19.

The next step is an important one. Think about it for a moment. He counts the number of nails he has taken, and places some of them back in the dish. The number he returns is determined by the number of nails he has selected. If he chose 11 nails he would return two nails. If you go through all of the possible numbers you will find that nine nails will always be left in your friend's hand. Here's how it works:

11 nails selected, 2 returned = 9 nails in hand
12 nails selected, 3 returned = 9 nails in hand
13 nails selected, 4 returned = 9 nails in hand
14 nails selected, 5 returned = 9 nails in hand
15 nails selected, 6 returned = 9 nails in hand
16 nails selected, 7 returned = 9 nails in hand
17 nails selected, 8 returned = 9 nails in hand
18 nails selected, 9 returned = 9 nails in hand
19 nails selected, 10 returned = 9 nails in hand

No matter what number your friend selects, after this step he will be left with nine nails in his hand. He can then remove, and hide, any number he likes. But once he hands you the remaining nails, all you must do is subtract that number from nine and you will know how many he is hiding.

Suppose he chose to keep all of his nails and he didn't give you any. Naturally you would know that he has all 9 nails hidden away, but, to convince him you can "talk" to nails you should still ask the nails remaining in the bowl how many he took. This will make your trick look like real magic.

If you have enjoyed playing with this mathematical trick, perhaps you would enjoy working out other formulas. Suppose you have 29 nails in the bowl and you asked your friend to remove any number over 20. You could do the trick in the same way, but this time your "magic number" would be 18 instead of 9.

A LAZY MAN'S TRICK

Here's a math-magic trick something like the last one. It uses numbers, but requires two friends and, most surprising of all, you never have to see any of the nails at all. In fact, it is called the Lazy Man's Trick because you can sit back in an easy chair, with your eyes closed if you like, and let your friends do all the work. You'll still surprise them both.

How Many Nails

Things Needed:
A bag containing
   60 or more nails
Two friends
An easy chair for you

Tell one of your friends that you are going to show him a trick. Let's call this friend number one. Because you're lazy your other friend is asked to act as your assistant. We'll call him number two. Now you can sit down in your easy chair, so you cannot see either one of your friends, and give them the following instructions.

A. Ask your assistant, number two, to take any number of nails (between 10 and 20) from the bag.

B Friend number one is now asked to take just twice the number of nails that number two did.

C. Now, ask your number two to give number one exactly four of his nails.

D. Finally, ask number one to count the number of nails that number two has left. Then ask number one to give number two twice that number from his own nails.

When he has done this, you are able to tell number one exactly how many nails he has left. You never peek or see how many nails either of your helpers have, and neither of them says anything to you at any time. In fact, when you finally announce the correct number, both of your friends will be surprised. How do you do it? The trick sounds like magic, but it is nothing more than mathematics, and it is simple. To solve it you must only remember the number that you asked number two to give number one in step number C. In this example, you asked him to give four nails. Simply multiply this number by three and you will determine the final number that your friend number one, will be left with. In this example it does not matter how many nails were taken at the start, $4 \times 3 = 12$, and your friend, number one, will be left holding 12 nails at the end.

Why does this work? If you study it a bit, and try it a few times, you may discover the mathematical reason. Here is one further clue that might aid you: You can use any number under 10 in place of the 4 in step C. No matter what number you use, just multiply that number by 3 to determine the final total.

If you use 2 ($2 \times 3 = 6$) friend number one will have 6 nails

If you use 8 ($8 \times 3 = 24$) friend number two will have 24 nails

Any number will work and, by changing numbers each time you may repeat the trick with your same friends many times. Although changing numbers makes it more difficult for your friends to figure out your trick, it should actually make it easier for you to understand why it works.

6

Final Fun-damentals

If you like mysteries and puzzles, there are many of them to be solved. This final chapter will introduce you to a variety of problems you can investigate with nails. You'll find some you can do as you read, and others will continue to challenge you long after you return this book to its shelf.

To begin, here is a short collection of *thinking mysteries*. You really don't have to do anything to solve them, except think!

- The average sized house contains about 40,000 nails. If you assume that each nail averages 1/10 of an ounce in weight, and you know that there are 16 ounces in one pound, how many pounds of nails are required to hold a house together?

- A man buries two pounds of nails in the ground for one year. One pound is steel nails, and the other pound is aluminum nails. When he digs them up he discovers the aluminum nails still weigh one pound, but the steel nails weigh only 3/4 of a pound. Try to explain this.

- A carpenter has two nails, one steel and one aluminum. They are both the same size, and they both look identical. How could the carpenter tell them apart? (I thought of four different ways, how many can you think of?)

FALLING FOR SCIENCE

Why do things fall down? People have thought about that puzzle for a long time. One of the first people to consider it seriously was the famous Greek philosopher Aristotle who lived about 2500 years ago.

Aristotle's explanation was very simple. He said that everything had a place where it belonged. Stones, for example, belonged on the ground. If you lifted one up, then let it go, it simply moved back to where it belonged. This was Aristotle's way of talking about *gravity*, a word that would not be invented until hundreds of years later.

Aristotle's explanation went further: he said that a bigger rock was heavier because it belonged on the ground more than a smaller rock. He said, therefore, if you drop a small and a large rock together, the larger one will always fall to earth faster because it belongs there more than the smaller one.

The easiest way to find out whether Aristotle is correct would be to pick up two rocks and try it. Surprisingly, in Aristotle's time nobody even bothered to try it as an investigation. In fact, nobody had ever heard of doing a science experiment!

If you've never really thought about this puzzle, here is an investigation you can do to start:

A Paper And Nail Puzzle

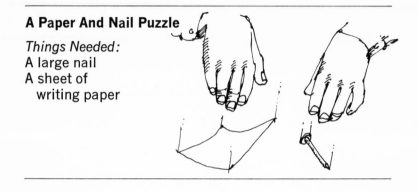

Things Needed:
A large nail
A sheet of
    writing paper

Hold the nail in one hand and the paper in the other. Hold them side by side, each the same height above the floor. Now release both of them at the same time. Which one strikes the earth first?

The nail weighs more than the sheet of paper. Is the nail like Aristotle's large rock, and the sheet of paper like the small one? Did the nail hit first because it is heavier? Was Aristotle correct?

About 2000 years after Aristotle died, another scientist also thought about falling objects. He was the Italian astronomer, Galileo. Galileo thought about, and considered, Aristotle's ideas, but he did not agree with them!

Galileo thought a big rock was made of a great many smaller ones lumped together. He thought, that if dropped, two rocks of the same size would fall at the same rate, and hit the ground together. If he held them very close they would still fall together, and at the same speed. In other words, he thought that all objects, regardless of their weight, would fall at the same speed.

Therefore, if you consider a big rock as being made of many small ones, then all small ones will drop at the same rate.

Galileo decided to do a very brave thing, he decided to actually *try* his ideas, rather than just think about them. He rolled different sized balls down ramps and observed how quickly they reached bottom.   Then, more convinced than ever that he was correct, he tried dropping some weights.

One of the most famous stories about Galileo suggests that he chose the Leaning Tower of Pisa to conduct his experiments.   This certainly would have been a wonderful choice, because of the tower's unusual angle.   Perhaps he did do it there, or perhaps that's only a story. Galileo, however, did drop some weights, and he did write his results down very carefully, and, what do you suppose he discovered?

Please try the investigation below out-of-doors on the lawn! Hold a hammer in one hand, and a nail in the other. Be sure they are both the same height above the ground.   Drop them at exactly the same time. Which one reached the ground first?   If you don't believe your results, try it several times.

Hammer and Nail In the Fall

Things Needed:
A hammer (any size)
A nail (any size)

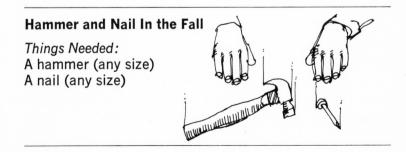

Galileo was right. All objects, regardless of their size or weight, fall at the same speed. The heavy hammer, and the lighter nail, reached the ground together.

If this doesn't sound right because of your earlier investigation, think about it. The nail appeared to fall faster than the sheet of paper because it really did.

This investigation was included for two reasons. First, to make you puzzle just a bit, and, second, to remind you of one of the reasons people are still confused by the way things fall. The paper would have fallen at exactly the same speed as the nail, except for the fact that it is surrounded by air. The air is under it; so, to fall downward the paper must first push the air out of the way. The nail must do this also but, because it is smaller it does not have to move as much air and can move downward faster.

Without air to confuse us, the paper and the nail would fall at the same rate. When you visit the moon perhaps you might like to conduct these same experiments. Without air to slow down the falling objects, all things on the moon should fall at exactly the same speed regardless of how big, small, heavy, light, fat, thin, tall, short—or anything else they might be.

NAILS THAT FLOAT AND SINK

Having been introduced to two famous scientists, Aristotle and Galileo, here is another one, Archimedes. He was a Greek scientist and inventor who lived more than 2000 years ago. There is a wonderful story about Archimedes that is worth re-telling.

Archimedes had been asked, by the king, to determine whether the royal crown was made of pure gold or a mixture of gold with silver. He was considering that problem while he was, of all things, taking a bath! When he was floating in the water, Archimedes noticed that he felt lighter and the water level was higher. "Eureka!" he shouted. For suddenly he was certain that he knew how to solve the king's mystery. Archimedes reasoned that the amount of water that rose from the original level would be equal to his own weight. He also knew that gold was a much denser metal than silver, or even a mixture of gold and silver. So, by immersing the crown in water, and carefully measuring the water that was displaced, he determined that the crown was less dense than it should have been if it were actually made of pure gold!

Archimedes discovered that any object placed in water will raise the level. This is called *displacement*. He also knew that the displacement would be different depending on whether the object was floating on the water, or sitting immersed, on the bottom. There is a puzzle you can do to investigate this difference:

A Boat Full of Nails

Things Needed:
A large bowl (or small sink)
  half-filled with water
A small soup dish
A bag of nails
A pencil

Float the soupbowl in the water in the larger bowl. With a pencil make a mark on the inside of the large bowl, right at the water line. Now, fill the soupbowl with nails. Put as many as you possibly can in, without sinking the boat. The nails will push down against the water and the water will be displaced upward. When you've finished loading the soupbowl, the amount of water that is above your original pencil mark would weigh exactly the same as the nails themselves. This is Archimedes principle. Make another pencil mark on the large bowl, at the new water line.

Now, suppose your boat tipped over and all the nails sank to the bottom? Would your waterline go up, down, or remain the same? That's an interesting question to ponder.

When the nails were in the boat, the water supported their entire weight. When the nails are sitting on the bottom of the bowl and the water is no longer support- ing their weight, the water line cannot return to its original mark because the nails are still in it.

When an object floats, it displaces its own weight of water, but, if it sinks another science rule comes into effect: The sunken object will displace only its own *volume* of water. Volume is the amount of space occupied by the object and it has nothing to do with the object's weight.

Why not carefully remove the nails from the floating soupbowl, and drop the nails into the water. Watch the waterline as you do. If you observe the results carefully you will be able to explain better the difference be- tween the science of things that float, and the science of things that sink.

THE WONDERFUL WEIGHING PUZZLES

Here is a collection of puzzles that both you, and your friends may enjoy solving. They require several objects of different weights. Nails are ideal.

To set up a weighing puzzle you will require a simple balance. You can make one from a coat hanger:

Building A Balance

Things Needed:
Coat hanger
Two paper clips
Sticky tape

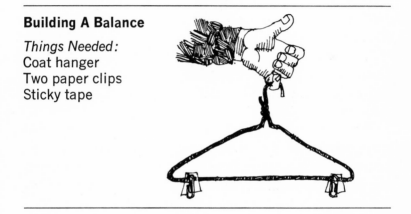

You do not actually have to weigh anything in a weighing puzzle, all you have to do is determine which of two items is the heavier. Loop the clips onto the hanger, and fasten them at the ends with sticky tape, as shown in the illustration. By hanging objects from the clips, and holding the coat hanger looped over your finger, you have a satisfactory *balance* that will do the job for you.

With this balance, and a few nails, you are ready to meet your first challenge. Let's make it an easy one:

A Weight Problem

Things Needed:
Coat hanger balance
Three large nails
One small nail
Four envelopes

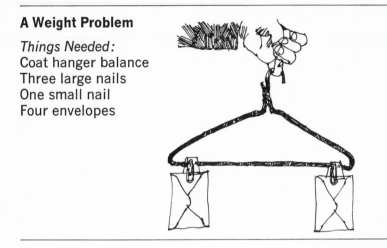

Place one of the nails in each envelope. Mix them together so you do not know which envelope contains the small nail. Here is the puzzle: You may fasten as many envelopes on either paper clip as you like. You must determine which envelope contains the light nail and, you can make no more than *two* weighings!

This elementary puzzle can actually be solved in two different ways. I'll explain one, and leave the other for you to discover.

Put one envelope on each side of the balance. If they do not balance, the side that goes up must be the light nail. If they do balance, they must both be heavy nails. Use the other two envelopes for your second weighing and watch for the light end.

This puzzle is actually too simple. However, other weighing puzzles are much more difficult. They become so when you add more nails:

Find One Nail In Six

Things Needed:
Coat hanger balance
Five large nails
One small nail
Six envelopes

Again seal the nails, one per envelope. Your problem, just as before, is to find the single light nail in just *two* weighings.

Naturally, before actually touching the envelopes, you should think about the best combinations to use. The envelopes you select should be a combination that will give you the greatest amount of information. Just as before, there are two different combinations you might use. Here's one; you'll have to think of the other way yourself:

Leave any two envelopes on the table. Clip the others, two on each side, to the balance . . .

. . . if the sides balance, the two remaining on the table contain the light nail. Use those two for your second weighing.

. . . if the sides do not balance, but one side goes up, that side contains the light nail. Attach the two envelopes, from the light side, to the scale for the second weighing.

I do not mean to "spoil your fun" by explaining how to solve the problems for you. There are still many, many more that you will have to work out yourself. You may use nails, envelopes, and the balance to investigate them ... or perhaps you can work them out right in your head.

Can you ...

1. Find 1 light from 2 heavy nails in just 1 weighing?
2. Find 1 light from 6 heavy nails in 2 weighings?
3. Find 2 light from 2 heavy nails in 2 weighings?
4. Find 1 light from 9 heavy nails in 3 weighings?
5. Find 3 light from 2 heavy nails in 2 weighings?
6. Find 6 light from 1 heavy nail in 2 weighings?

I couldn't find any way to find ... 1 light from 3 heavy nails in just 1 weighing ... can you?

Perhaps, also, when you have solved all of these puzzles, you might enjoy the challenge of making up new ones. You'll find that is even harder than solving them.

THE SHADOW OF A NAIL

How high is your house? How tall is a tree? You can easily find out using a nail, its shadow, and some simple mathematics.

When the sun shines, everything in the path of its light casts a shadow picture of itself on the ground. This shadow may be long or short, depending on where the sun is in the sky. But there is always a *ratio* between the height of the object and the length of its shadow. This ratio can help you tell the height of any object that is impossible to measure with a ruler.

Here's how it works:

Suppose it's around one o'clock, you are four feet tall, and you cast a shadow two feet long. You and your shadow, would have a ratio of 4/2. This indicates that you are exactly twice as long as your shadow.

If your father is six feet tall, and he measured his shadow at the same time you did, he would find his shadow 3 feet long. His ratio, 6/3, indicates the same thing as yours. It shows that he is twice as tall as his shadow is long.

Now, suppose, knowing that the shadow ratio was 2/1 (objects are twice as tall as their shadows), you measured the length of the shadow of a tree in your yard and found it was ten feet long. At the moment you do not know how tall the tree is, so let's call this height H for the unknown. The tree's shadow ratio would be H/10. How would you find the number represented by H?

Just compare the tree with yourself!

You already know your ratio because you know how tall you are. You can write down what you know about yourself, and the tree as a mathematical ratio:

$$\text{(You)} \quad \frac{4}{2} = \frac{H}{10} \quad \text{(Tree)}$$

This ratio shows that your height casts a shadow of a certain size (two feet), and that the tree's shadow is ten feet. The same ratio must exist for the heights of both you and the tree that exists for your shadow.

To find the height of the tree, you need only to do the following:

$$\frac{4}{2} = \frac{H}{10}$$

Multiply the numerator (the top number) of one ratio by the denominator (the bottom number) of the other. Or, 2 × H and 4 × 10. This gives

2H = 40

Now, divide the number on the left into the number on the right:

H = 2)‾40‾
so, H = 20 feet (the height of the tree)

To prove that this is correct, and to show that the ratios between shadow length and height are equal, you might solve the problem again, using 20 feet instead of H:

$$\frac{4}{2} = \frac{20}{10}$$

Again, multiply numerators with denominators and you get:

$$\frac{40}{40}$$ . . . which shows that the ratios are equal!

Once you understand this simple process you can easily make a device which will enable you to determine the height of any object that is casting a shadow. This trick for calculating height is called *shadow reckoning*. It may prove useful the next time you are trying to guess how tall a tree, water tower, radio antenna, or building is.

Nail Yourself a Shadow Meter

Things Needed:
A long nail ($3\frac{1}{2}''$ to $4''$)
A board about 13" long
A one-foot ruler
A spool of string
Sticky tape

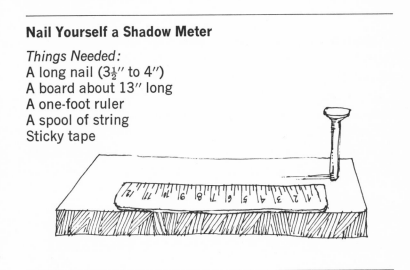

Pound the nail into the board at one end, as shown in the illustration. Drive the nail in carefully, so that exactly three inches of nail protrudes above the board. Measure this with a ruler to be precise. Now lay the ruler along the board, starting with the end right next to the nail and fasten it in position with the sticky tape. Be sure that your nail is pointing straight up from the board and your shadow meter is ready to go.

Lay the shadow meter on the ground, right beside the shadow you would like to measure. Turn the board so the nail's shadow falls along the board right beside the ruler. Make a note of how long the nail's shadow is. Also make a note that the nail itself is 3 inches high. Now, you will be using this formula:

$$\frac{\text{Height of the nail}}{\text{length of nail's shadow}} = \frac{\text{height of object}}{\text{length of object's shadow}}$$

You can fill in the numbers and work out the formula just as you did before. You will have to know the length of the object's shadow, and you can get this by stretching a string down its length and using your ruler to measure the string. As an example, let's assume that the nail casts a nine inch shadow, and the object casts a 30 foot shadow. Filling these numbers into the formula would make it appear like this:

$$\frac{3 \text{ inches}}{9 \text{ inches}} = \frac{\text{height of the object (H)}}{30 \text{ feet}}$$

Then, multiplying numerators with denominators:

$9H = 90$

So,

$H = 10$ feet (height of the object)

In this example, the shadows were *long*, as they would be in early morning or late in the afternoon. Your shadow meter will also work well in mid-afternoon, when the sun is high in the sky and shadows are *short*. If you try it at a time when the shadow of the nail is longer than 12 inches you will not be able to get a proper measurement. Simply select another time of day and you will be successful.

ON YOUR OWN

Finally, here is a collection of odd puzzles for you to think about by yourself. You should be able to solve them all with a little careful looking and thinking. Perhaps they will lead you to further questions. I hope so.

- A nail, with a head, can be used to trace a large circle on the floor. Can you explain how? If not, try rolling one.

- You can easily arrange four nails to make the following patterns: a straight line, a square, a parallelogram, a diamond and a cross. But these are not the only possible patterns. How many others can you discover?

- Here's a wonderful puzzle to play with: Lay a nail, with a head, flat on the table. Now, let's see you pick up that nail using only your index finger.

- Most wooden things are held together with nails, and yet you seldom see nails showing. Look carefully around your house. How many ways can you discover that a carpenter hides nails so they don't show?

- Look carefully at a nail. The head makes a circle and the sides make a straight line. Can you find the shape of a diamond, or a square, on the nail?

- Here is a tricky problem: Can you drop a nail, with a head, on the table in such a way that it will always land on its head with the point sticking straight up? Here's a clue: in addition to the nail you will need a sheet of paper.

- When a nail is dropped into a glass dish it makes a sound that is quite musical. If a small nail and a large nail were dropped, one after the other, which one do you suspect would make the higher note?

- Which is harder, a nail or a piece of glass? Try scratching an old jar or bottle to find out if you're right.

Perhaps, by now, you would like to invent some science demonstrations using nails by yourself. Here's a good way to start: there are many different kinds, sizes, and shapes of nails you can find in a hardware store. Perhaps one of these special nails could be used in a special investigation. Look around a hardware store and you will find nails with two heads, nails with off-center heads, nails that are covered with a layer of zinc, nails almost a foot long and tiny tacks that are sterilized and germ-free. Most of these nails have special names too. See if you can learn their names, find out what they do, and then invent your own special investigations.

Index

107

ABOUT THE AUTHOR

Besides his ability to use an ordinary nail to explain the mystery of magnetism, Laurence B. White, Jr. is also a master builder of paper architecture and a serious collector of rubber bands. Yet, these seemingly unrelated activities all reflect his principal interest: the exciting presentation of scientific principles. As Assistant Director of the Needham (Massachusetts) Elementary Science Center, as a popular television teacher and as a member in good standing of the Society of American Magicians, Mr. White has had ample opportunities to display his talents.